hands-on
maths

Kerry Dalton

Year 6

Published by Keen Kite Books
An imprint of HarperCollins*Publishers* Ltd
The News Building
1 London Bridge Street
London
SE1 9GF

ISBN 9780008267001

First published in 2018
10 9 8 7 6 5 4 3 2 1

Text and design © 2018 Keen Kite Books, an imprint of HarperCollins*Publishers* Ltd
Author: Kerry Dalton

Series Concept and Commissioning: Shelley Teasdale and Michelle I'Anson
Project Manager: Fiona Watson
Editor: Denise Moulton
Cover Design: Anthony Godber
Text Design and Layout: Contentra Technologies
Production: Natalia Rebow

A CIP record of this book is available from the British Library.

Contents

Year 6 aims and objectives   4

Introduction   6

Resources   7

Assessment, cross-curricular links and vocabulary   8

Ratio and proportion
Week 1   9
Week 2   10
Week 3   11
Week 4   12
Week 5   13
Week 6   14

Place value
Week 1   15
Week 2   16
Week 3   17
Week 4   18
Week 5   19
Week 6   20

Algebra
Week 1   21
Week 2   22
Week 3   23
Week 4   24
Week 5   25
Week 6   26

Addition and subtraction / Fractions
Week 1   27
Week 2   28
Week 3   29
Week 4   30
Week 5   31
Week 6   32

Multiplication and division / Fractions
Week 1   33
Week 2   34
Week 3   35
Week 4   36
Week 5   37
Week 6   38

Fractions (including decimals and percentages)
Week 1   39
Week 2   40
Week 3   41
Week 4   42
Week 5   43
Week 6   44

Vocabulary cards   45

Isometric dotty paper   46

Digit cards   47

Year 6 aims and objectives

Hands-on maths Year 6 encourages pupils to understand and master a range of mathematical concepts, through a practical and hands-on approach. Using a range of everyday objects and common mathematical resources, pupils will explore and represent key mathematical concepts. These concepts are linked directly to the National Curriculum 2014 objectives for Year 6. Each objective will be investigated over the course of the week using a wide range of hands-on approaches such as using Dienes, place-value counters, playing cards, dice, place-value grids, practical problems and a mix of individual and paired work. The mathematical concepts are explored in a variety of contexts to give pupils a richer and deeper learning experience, supporting a mastery approach.

The National Curriculum 2014 aims to ensure that, in upper Key Stage 2, pupils can extend their understanding of the number system and place value to include fractions, decimals and percentages in detail. This should develop the connections that pupils make between core concepts such as multiplication and division with numbers less than one, and decimals, percentages, ratio and proportion.

In Years 1 to 5, we explored counting as the first module. However, as ratio and proportion are first explored in Year 6, this has replaced the counting module as the first module in Year 6. It also ensures that the same topics are maintained across the whole school.

Year 6 programme and overview of objectives

Topic	Week 1	Week 2	Week 3	Week 4	Week 5	Week 6
Ratio and proportion	Solve problems involving the relative sizes of two quantities where missing values can be found by using integer multiplication facts	Solve problems involving the relative sizes of two quantities where missing values can be found by using integer division facts	Solve problems involving the calculation of percentages	Solve problems involving similar shapes where the scale factor is known or can be found	Solve problems involving unequal sharing and grouping using knowledge of fractions	Solve problems involving unequal sharing and grouping using knowledge of multiples
Place value	Read and write numbers up to 10 000 000 and determine the value of each digit	Order and compare numbers up to 10 000 000 and determine the value of each digit	Compare numbers up to 10 000 000 and determine the value of each digit	Round any whole number to a required degree of accuracy	Use negative numbers in context, and calculate intervals across zero	Solve number and practical problems that involve place value

Year 6 aims and objectives

Topic	Week 1	Week 2	Week 3	Week 4	Week 5	Week 6
Algebra	Use simple formulae	Generate linear number sequences	Describe linear number sequences	Express missing number problems algebraically	Find pairs of numbers that satisfy an equation with two unknowns	Enumerate possibilities of combinations of two variables
Addition and subtraction / Fractions	Perform mental calculations, including with large numbers	Add fractions with different denominators using the concept of equivalent fractions	Add fractions with mixed numbers, using the concept of equivalent fractions	Subtract fractions with different denominators using the concept of equivalent fractions	Subtract fractions with mixed numbers, using the concept of equivalent fractions	Perform mental calculations, including with mixed operations
Multiplication and division / Fractions	Perform mental calculations, including with mixed operations and large numbers	Identify common factors	Identify common multiples	Identify prime numbers	Multiply simple pairs of proper fractions, writing the answer in its simplest form	Divide proper fractions by whole numbers
Fractions (including decimals and percentages)	Use common multiples to express fractions in the same denomination	Use common factors to simplify fractions	Identify the value of each digit in numbers given to three decimal places	Multiply numbers by 10, 100 and 1000 giving answers up to three decimal places	Divide numbers by 10, 100 and 1000 giving answers up to three decimal places	Multiply one-digit numbers with up to two decimal places by whole numbers

Introduction

The *Hands-on maths* series of books aims to develop the use of readily available manipulatives such as toy cars, shells and counters to support understanding in maths. The series supports a concrete–pictorial–abstract approach to help develop pupils' mastery of key National Curriculum objectives.

This title covers six topic areas from the National Curriculum (ratio and proportion; place value; algebra; the four number operations: addition and subtraction and multiplication and division; and fractions). Each area is covered during a six-week unit, with an easy-to-implement 10-minute activity provided for each day of the week. Photos are included for each activity to support delivery.

Hands-on maths enables a deep interrogation of the curriculum objectives, using a broad range of approaches and resources. It is not intended that schools purchase additional or specialist equipment to deliver the sessions; in fact, it is hoped that pupils will very much help to prepare resources for the different units, using a range of natural, formal and typical maths resources found in most classrooms and schools. This will help pupils to find ways to independently gain a deep understanding and enjoyment of maths.

A typical 'hands-on' classroom will have a good range of resources, both formal and informal. These may include counters, playing cards, coins, Dienes, dominoes, small objects such as toy cars and animals, Cuisenaire rods, 100 squares and hoops.

There is no requirement to use *only* the resources seen in the photographs that accompany each activity. Cubes may look like those in the green bowl, or will be just as effective if they look like the ones in the blue bowl. They serve the same purpose in helping pupils understand what the cubes represent.

Resources

Hands-on maths uses a range of formal, informal and 'typical' resources found in most classrooms and schools. To complete the activities in this book, it is expected that teachers will have the following resources readily available:

- whiteboards and pens for individual pupils and pairs of pupils
- Dienes and Cuisenaire rods
- dice, coins and bead strings
- a range of cards, including playing cards, place-value arrow cards and digit cards

- collections of objects that pupils are interested in and want to count, such as toy cars, toy animals and shells
- bowls / containers to store sets of resources in, making it easy for pupils to handle and use the objects

- ten frames (these could be egg boxes, ice-cube trays, printed frames or something pupils have created themselves)
- number lines and 100 squares – lots of different types and styles: printed, home-made, interactive, digital or practical … whatever you prefer, and whatever is handy. (For 100 squares, there is, of course, the 1–100 or 0–99 choice to make; both work and it is best to choose whatever works for the class. Both offer a slight difference in place-value perspective, with 0–99 giving the 'zero as a place holder' emphasis, while the 1–100 version helps pupils to visualise the position of 100 in relation to the two-digit numbers.)

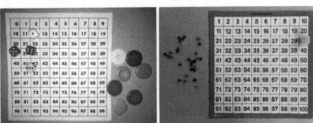

- counters and cubes – lots of them! Many of the activities require counters and cubes to be readily available. The cubes can be any size and any colour: what the cubes represent is the most important factor.

Maths is a truly unique, creative and exciting discipline that can provide pupils with the opportunity to delve deeply into core concepts. Maths is found all around us, every day, in many different forms. It complements the principles of science, technology and engineering.

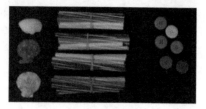

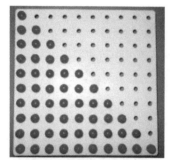

Hands-on maths provides ideas that can be adapted to suit the broad range of needs in our classrooms today. These ideas can be used as a starting point for assessment – before, during or after teaching of a particular topic has taken place. The activities are intended to be flexible enough to be used with a whole class and can, of course, be differentiated to suit individual pupils in a class.

The activities can be adapted to link to other subject areas and interests. For example, a suggestion to use farm animals may link well to a science unit on classification or food chains; alternatively, the resource could be substituted with bugs if minibeasts is an area of interest for pupils. Teachers can be as flexible as they wish with the activities and resources – class teachers know their pupils best.

Spoken language is underpinned in maths by the unique mathematical vocabulary pupils need to be able to use fluently in order to demonstrate their reasoning skills and show mathematical proof. The correct, regular and secure use of mathematical language is key to pupils' understanding; it is the way in which they reason verbally, negotiate conceptual understanding and build secure foundations for a love of mathematics and all that it brings. Each unit in *Hands-on maths* identifies a range of vocabulary that is typical of, but by no means limited to, that particular unit. The way the vocabulary is used and incorporated into activities is down to individual style and preference and, as with all of the resources in the book, will be very much dependent on the needs of each individual class. A blank template for creating vocabulary cards is included at the back of this book.

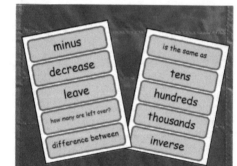

Week 1: Ratio and proportion

Solve problems involving the relative sizes of two quantities where missing values can be found by using integer multiplication facts

Resources: cubes, whiteboards and pens

Vocabulary: divisible (by), divisibility, factor, factorise, part, equal parts, fraction, proper / improper fraction, mixed number, numerator, denominator, equivalent, reduced to, simplify, one whole, half, quarter, eighth, third, sixth, ninth ... hundredth, thousandth, proportion, ratio, in / for / to every, as many as, decimal, decimal fraction / point / place, percentage, per cent, %

Monday

Introduce *ratio* as a concept, i.e. explain that a ratio shows the relative sizes of two or more quantities of the same unit. Ratios can be shown in different ways – either using the colon symbol to separate values, or as a single number by dividing one value by the total and expressing it as a fraction, a decimal or a percentage.

Using cubes, demonstrate that in the ratio 1:3 there is 1 red for every 3 blue cubes. $\frac{1}{4}$ are red and $\frac{3}{4}$ are blue. 0.25 are red (by dividing 1 by 4) and 0.75 are blue. 25% are red and 75% are blue. Ask pupils to copy your model and write the ratios of the colours, as shown, using 4 cubes (1 of one colour and 3 of another colour) and whiteboards.

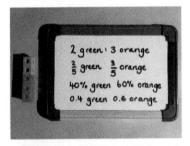

Tuesday

Give each pair of pupils some cubes of different colours.

Working in pairs, pupils explore ways of comparing one part of a whole to another. Partner 1 in each pair creates a simple cube tower in two colours and partner 2 writes the ways of comparing part to part (e.g. 2:3) and part to whole (e.g. $\frac{2}{5}$). Repeat with pupils swapping roles.

Wednesday

Prepare some cube towers and ask pupils to express these as a comparison of one part to the other. Encourage simplification where this is possible.

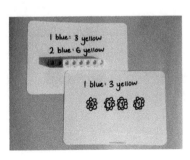

Thursday

Give each pair of pupils some cubes of different colours.

Using cubes and diagrams, ask pupils to represent the following: 'A gardener is creating coloured flower displays for a special event. He needs to plant 3 yellow plants for every 1 blue plant.' Partner 1 in each pair uses cubes and partner 2 draws a diagram to represent the ratios. Ask: 'How many yellow plants will he need to plant if he plants 3 blue plants, then 5, 10, 20, etc.?'

Friday

Repeat Thursday's activity, with pupils swapping roles.

Week 2: Ratio and proportion

Solve problems involving the relative sizes of two quantities where missing values can be found by using integer division facts

Resources: cubes, whiteboards and pens

Vocabulary: divisible (by), divisibility, factor, factorise, part, equal parts, fraction, proper / improper fraction, mixed number, numerator, denominator, equivalent, reduced to, simplify, one whole, half, quarter, eighth, third, sixth, ninth ... hundredth, thousandth, proportion, ratio, in / for / to every, as many as, decimal, decimal fraction / point / place, percentage, per cent, %

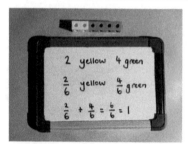

Monday

Introduce *proportion* as a concept, encouraging discussions about proportion being a comparison in relation to a whole amount, whereas ratio compares one part of a whole amount to another part of the same whole amount.

Make a tower of 2 yellow and 4 green cubes. Demonstrate that 2 out of 6 are yellow and 4 out of 6 are green, and together they make the whole amount. Ask pupils to copy your model, using 6 cubes and whiteboards.

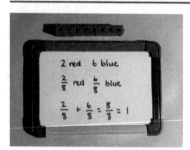

Tuesday

Give each pair of pupils some cubes of different colours.

Working in pairs, pupils explore ways of expressing proportion with 8 cubes (2 of one colour and 6 of another).

Encourage simplification to $\frac{1}{4}$ and $\frac{3}{4}$ respectively.

Wednesday

Give each pair of pupils some cubes of different colours.

Using cubes and written fractions, ask pupils to represent the following: 'A football team is selling home and away kits. The home kit is red and the away kit is white. After a week, the club had sold 100 kits. 70 were home kits and 30 were away kits. How would you represent this as a fraction?' Partner 1 in each pair uses cubes and partner 2 writes the fraction on their whiteboard. Ask: 'What fraction of the kits were home and away in the following proportions: 40 and 60, 80 and 20, 35 and 65, etc.?'

Thursday

Repeat Wednesday's activity, with pupils swapping roles.

Friday

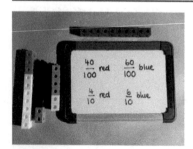

Prepare some cube towers and ask pupils to express these as a comparison of one part to the whole amount. Tell pupils that each cube represents 10 football kits.

Encourage simplification where this is possible.

Week 3: Ratio and proportion

Solve problems involving the calculation of percentages

Resources: cubes, whiteboards and pens

> **Vocabulary:** divisible (by), divisibility, factor, factorise, part, equal parts, fraction, proper / improper fraction, mixed number, numerator, denominator, equivalent, reduced to, simplify, one whole, half, quarter, eighth, third, sixth, ninth ... hundredth, thousandth, proportion, ratio, in / for / to every, as many as, decimal, decimal fraction / point / place, percentage, per cent, %

Monday

Tell pupils that you have been given a bank statement, but taxes at 50% have been deducted and you can only see the remaining 50% of the balance. Model using a bar model made of cubes to represent the monies. Each cube represents £10 000.

Using whiteboards, give pupils other balances and ask them to calculate the original amounts.

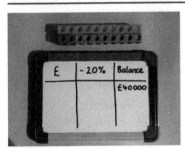

Tuesday

Tell pupils that you have been given another bank statement, but today taxes at 20% have been deducted and you can only see the remaining 80% of the balance. Model using a bar model made of cubes to represent the monies. Each cube represents £5000.

Using whiteboards, give pupils other balances and ask them to calculate the original amounts.

Wednesday

Tell pupils that you have been given another bank statement, but today you can only see the tax that was deducted. Tax was charged at 10%. Use a bar model made of cubes to represent the monies. Each cube represents £100 000.

Using whiteboards, give pupils other balances and ask them to calculate the original amounts.

Thursday

Tell pupils that you have been asked to check the prices for a bike shop, but have only been given the sale prices. These are the original selling prices with 10% already deducted.

Write the amounts on the board. Working in pairs, ask partner 1 in each pair to calculate the original selling prices while partner 2 calculates the saving.

Ask pupils to check their answers by adding the sale price to the saving and checking the total against the original selling price.

Friday

Repeat Thursday's activity, with pupils swapping roles.

Week 4: Ratio and proportion

Solve problems involving similar shapes where the scale factor is known or can be found

Resources: squared paper, whiteboards and pens, isometric dotty paper

Vocabulary: divisible (by), divisibility, factor, factorise, part, equal parts, fraction, proper / improper fraction, mixed number, numerator, denominator, equivalent, reduced to, simplify, one whole, half, quarter, eighth, third, sixth, ninth ... hundredth, thousandth, proportion, ratio, in / for / to every, as many as, decimal, decimal fraction / point / place, percentage, per cent, %

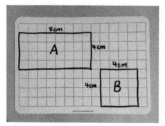

Monday

Using squared paper or whiteboards, draw the rectangles shown. Explain that rectangle B is half as wide as rectangle A.

Ask pupils to draw rectangles A and B, filling in the missing lengths. Ask pupils to say what is similar and what is different about the diagrams.

Repeat with rectangles 8cm × 3cm, 9cm × 4cm, 12cm × 5cm.

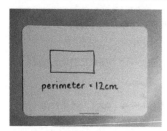

Tuesday

Explain that you have drawn a rectangle with a perimeter of 12cm on your board and that you would like pupils to draw a rectangle with twice the perimeter. How many different possibilities are there? (Rectangles with sides 10cm × 2cm, 8cm × 4cm, 9cm × 3cm, 7cm x 5cm and 11cm × 1cm would all give a perimeter twice 12cm, i.e. 24cm.)

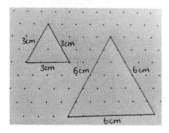

Wednesday

Using isometric dotty paper (see the back of the book), draw an equilateral triangle with a perimeter of 18cm. Ask pupils to draw their own triangle and to explain how to calculate the length of each side (18cm ÷ 3). Now ask pupils to create an equilateral triangle that has half the perimeter. Ask pupils to say what is similar and what is different about the diagrams.

Repeat with equilateral triangles with perimeters of 24cm and 30cm.

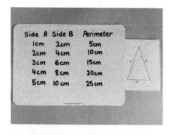

Thursday

Explain that you have drawn an isosceles triangle on your dotty paper; the length of the longest sides (B) is twice as long as the length of the shortest side (A). In total the perimeter is less than 30cm.

What are all the possible lengths of sides A and B if the side lengths are whole numbers only? (Answers as shown.)

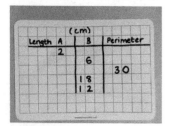

Friday

Draw the incomplete table shown on the board.

If the long sides (B) of a rectangle are double the length of the short sides (A), and the total perimeter is less than 80cm, what are the possible lengths if the side lengths are whole numbers only? Ask pupils to fill in the missing numbers.

Week 5: Ratio and proportion

Solve problems involving unequal sharing and grouping using knowledge of fractions

Resources: cubes, bowls, whiteboards and pens, counters / beads

Vocabulary: divisible (by), divisibility, factor, factorise, part, equal parts, fraction, proper / improper fraction, mixed number, numerator, denominator, equivalent, reduced to, simplify, one whole, half, quarter, eighth, third, sixth, ninth … hundredth, thousandth, proportion, ratio, in / for / to every, as many as, decimal, decimal fraction / point / place, percentage, per cent, %

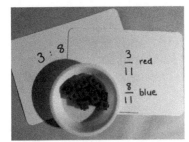

Monday

Prepare enough bowls of cubes of two different colours (between 10 and 20 cubes in each bowl) for pupils to work in pairs. Place the bowls around the room, labelled A, B, C, D, etc.

Working in pairs, ask partner 1 in each pair to write the ratio of the cubes in the bowl and partner 2 to write the proportion of the cubes on their whiteboards.

Tuesday

Repeat Monday's activity, with pupils swapping roles.

Wednesday

Give each pupil a set of counters / beads – each pupil will need approximately 30 counters / beads.

Explain that pupils will be making designs using repeating patterns. Ask pupils to create a design that uses colours in the ratio 2:4 with a repeating pattern.

Thursday

Use the counters / beads from Wednesday's activity.

Explain that pupils are again making designs using repeating patterns. Ask pupils to create a design that uses 1 of one colour for every 3 of a second colour. They must keep an accurate repeating pattern. Can they create a design with 21 counters / beads? (No) Why not? (Because the pattern uses 4 counters / beads in every repeat.)

Friday

Give each pair of pupils 48 counters / beads.

Explain that pupils are again making designs using repeating patterns. Ask pairs of pupils to create a design that uses 1 of one colour for every 4 of a second colour. They must keep an accurate repeating pattern. They can use a maximum of 48 counters / beads per design. What is the most and least number of counters / beads in each design? (Least is 10 and most is 45, as each repeat uses 5 counters / beads.)

Week 6: Ratio and proportion

Solve problems involving unequal sharing and grouping using knowledge of multiples

Resources: whiteboards and pens

> **Vocabulary:** divisible (by), divisibility, factor, factorise, part, equal parts, fraction, proper / improper fraction, mixed number, numerator, denominator, equivalent, reduced to, simplify, one whole, half, quarter, eighth, third, sixth, ninth ... hundredth, thousandth, proportion, ratio, in / for / to every, as many as, decimal, decimal fraction / point / place, percentage, per cent, %

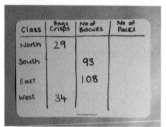

Monday

Explain that, this week, pupils will be planning parties.

Draw the incomplete table shown and ask pupils to fill in the missing values. The completed table will show the correct number of knives, forks and spoons needed for the party, based on pupil numbers today. As well as a knife and fork, each guest will need two spoons.

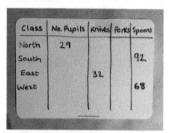

Tuesday

Draw the incomplete table shown and ask pupils to calculate how many bags of crisps, how many biscuits and how many packs of biscuits each class will need.

The completed table will show the correct number of bags of crisps and packets of biscuits needed for the party, based on pupil numbers today. Each pupil gets one bag of crisps and three biscuits. Biscuits come in packs of 30.

Wednesday

Explain that each packet of lollies costs 60p (with 12 lollies in each pack) and each pack of sausage rolls costs £1.20 (with 6 sausage rolls in each pack).

Tell pupils that you have a budget of £40 for lollies and sausage rolls. What combination(s) of lollies and sausage rolls could be bought based on 120 pupils?

Thursday

Draw the incomplete table shown and ask pupils to calculate how many of each type of sandwich they will need to make. Pupils must make the correct number of sandwiches, based on 120 pupils. You have already surveyed pupils to find their preferred fillings. Each pupil gets four small sandwiches. Ask pupils to keep their calculations for Friday. (If pupils are confident, this could be extended to explore the number of slices of bread, mass of fillings, etc.)

Friday

On the party day, only 90% of each type of sandwich are eaten. They come in trays of 12 sandwiches. How many complete trays are wasted in total?

Week 1: Place value

Read and write numbers up to 10 000 000 and determine the value of each digit

Resources: whiteboards and pens, cubes / counters, timer

Vocabulary: place value, place, ones, tens, hundreds, thousands, millions, digit, one-, two- … seven-digit number, tenths, hundredths, thousandths, represents, the same as, equal to, greater, more, larger, less, fewer, smaller, greatest, most, largest, least, fewest, smallest, one … ten … hundred … thousand more or less, compare, order, first, second, third … last, numeral, consecutive

Monday

Write 2, 4, 7, 3, 6, 8 on the board. Demonstrate how these digits can be used to create different numbers based on their position. Say the numbers aloud together.

Ask pupils to use these digits to make as many six-digit numbers as possible in 3 minutes. (There are hundreds of possibilities.) Go round the group, asking pupils to each say a number from their whiteboard, with pupils ticking when one of their numbers is read aloud.

Tuesday

Repeat Monday's activity, using the digits 3, 1, 4, 5, 6, 8, 9 and asking pupils to make seven-digit numbers.

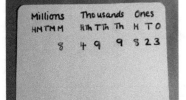

Wednesday

Ask pupils to draw a place-value grid, as shown, on their whiteboards.

Read out criteria to create a seven-digit number and then ask pupils to read their number aloud. For example:

- an even digit in the millions column
- 4 in the hundred thousands column
- 8 in the hundreds columns
- the digit 9 so that it represents 9000
- a digit greater than 7 in the ten thousands column
- an odd number in the ones column
- an even number in the tens column.

Thursday

Give half of the group a whiteboard and pen each. Ask these pupils to write a seven-digit number in large writing on their whiteboard. These pupils then stand dotted around the room. (This works well in the hall or playground.)

The other half of the group are 'readers' and have 5 minutes to visit as many whiteboards as possible and read the number aloud. If they read it correctly, they receive a cube / counter. The winner is the pupil with the most cubes / counters at the end of the five minutes.

Friday

Repeat Thursday's activity, with pupils swapping roles.

Week 2: Place value

Order and compare numbers up to 10 000 000 and determine the value of each digit

Resources: blank cards / sticky notes, timers, whiteboards and pens

Vocabulary: place value, place, ones, tens, hundreds, thousands, millions, digit, one-, two- … seven-digit number, tenths, hundredths, thousandths, represents, the same as, equal to, greater, more, larger, less, fewer, smaller, greatest, most, largest, least, fewest, smallest, one … ten … hundred … thousand more or less, compare, order, first, second, third … last, numeral, consecutive

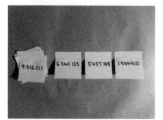

Monday

Give each pair of pupils ten blank cards or sticky notes and a timer.

Each pupil takes half of the cards / sticky notes and writes a seven-digit number on each. Partner 1 in each pair places the cards / sticky notes face down. Once partner 2 starts the timer, partner 1 has one minute to turn the cards over and place the numbers in order from smallest to largest. Partner 2 checks the order.

Swap partners with another pair so that partner 1 has to order a different set of numbers.

Tuesday

Repeat Monday's activity, with pupils swapping roles.

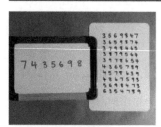

Wednesday

Write 7, 4, 3, 5, 6, 9, 8 on the board.

Ask pupils to generate ten different seven-digit numbers on their whiteboards, using these digits, and then to order the numbers from smallest to largest.

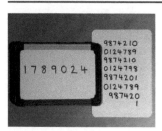

Thursday

Write 1, 7, 8, 9, 0, 2, 4 on the board. Using only the digits shown, ask pupils to write:

- the highest and lowest numbers
- the highest and lowest even numbers
- the highest and lowest odd numbers.

Challenge: using some of the digits, find the highest and lowest numbers that are a multiple of 3 (check! a number is divisible by of 3 if the sum of its digits is divisible by 3).

Friday

Ask pupils to divide their whiteboards into five parts as shown. Say a seven-digit number and ask pupils to write it in numerals at the top of their whiteboards. Pupils write other seven-digit numbers that are smaller than this number and that fit specific criteria, for example:

- 8 million and an even ones value
- an even number with 612 thousands and zero in the hundreds
- more than 300 thousands and a ones value that is less than the tens value
- 309 thousands and an odd tens value.

Week 3: Place value

Compare numbers up to 10 000 000 and determine the value of each digit

Resources: playing cards, whiteboards and pens, counters

Vocabulary: place value, place, ones, tens, hundreds, thousands, millions, digit, one-, two- … seven-digit number, tenths, hundredths, thousandths, represents, the same as, equal to, greater, more, larger, less, fewer, smaller, greatest, most, largest, least, fewest, smallest, one … ten … hundred … thousand more or less, compare, order, first, second, third … last, numeral, consecutive

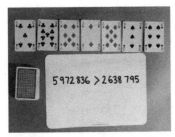

Monday

Take seven playing cards – remove the picture cards and tens (or pre-agree their value, e.g. picture cards have a value of 1 and tens have a value of 0).

Place the cards on the board or show on a visualiser. Call out an instruction, either 'greater than' or 'less than'. Pupils rearrange the digits shown on the cards and write an inequality that uses the correct symbol and the original number.

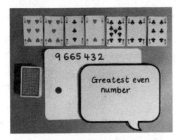

Tuesday

Place seven playing cards on the board or show on a visualiser.

Call out criteria; pupils use the digits on the cards to create a seven-digit number that meets the criteria, such as greatest even number, smallest odd number, largest multiple of 5. Pupils with a correct value win a counter.

Wednesday

Call out a digit from 1–9 and a place value. Pupils write a number on their whiteboards that fits the criteria and underline the digit you called out. Working in pairs, pupils swap whiteboards and say their partner's number aloud. Repeat.

Thursday

Call out a seven-digit number and ask pupils to write it on their whiteboards.

Tell pupils that, when you call out a place value, they should write a different number underneath that contains the same digit in the place value position you called out. Ask pupils to show you their whiteboards, checking for their place value accuracy.

Friday

Ask pupils to make a seven-digit number that has a digit total of 15.

- How many numbers can they create that have a digit total of 15?
- What is the smallest number they can create?
- What is the largest number they can create?

Week 4: Place value

Round any whole number to a required degree of accuracy

Resources: cubes, matchsticks, whiteboards and pens

Vocabulary: place value, place, ones, tens, hundreds, thousands, millions, digit, one-, two- … seven-digit number, tenths, hundredths, thousandths, represents, the same as, equal to, greater, more, larger, less, fewer, smaller, greatest, most, largest, least, fewest, smallest, one … ten … hundred … thousand more or less, compare, order, first, second, third … last, numeral, consecutive

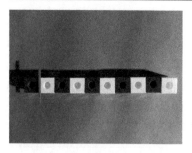

Monday

Give each pupil a matchstick and a stick of ten cubes to create a counting stick.

Tell them that one end of the counting stick represents 10 100 and the other end represents 10 200. Ask what each interval is worth. (10)

Now tell pupils that they should place the matchstick on 10 110 on their counting stick. Ask whether 10 110 would round to the nearest hundred to 10 200 or to 10 100. Repeat for 10 185, 10 010, 10 190, 10 175, 10 101.

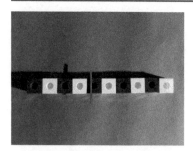

Tuesday

Using the same counting sticks, tell pupils that one end of the counting stick represents 14 000 and the other end represents 15 000. Ask the value of each interval. (100)

Now tell pupils to place the matchstick on 14 420 on their counting stick. Ask whether 14 420 would round up to the nearest hundred (14 500) or down (14 400). Repeat for 14 110, 14 360, 14 900, 14 750 and 14 404.

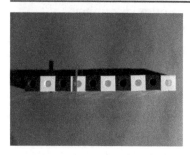

Wednesday

Using the same counting sticks, tell pupils that one end of the counting stick represents 30 000 and the other end represents 40 000. Ask the value of each interval. (1000)

Now tell pupils to place the matchstick on 33 200 on their counting stick. Ask whether 33 200 would round up to the nearest thousand (34 000) or down (33 000). Repeat for 36 900, 35 700, 31 040, 30 300 and 33 470.

Thursday

Using whiteboards, ask pupils to draw a target as shown, with 1000 in the centre.

Ask pupils to write numbers in the inner circle that, when rounded to the nearest hundred, would give the answer 1000. In the outer circle they write numbers that, when rounded to the nearest thousand, would also give the answer 1000.

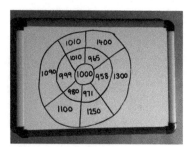

Friday

Repeat Thursday's activity, but with 10 000 written in the centre of the target and rounding numbers to the nearest thousand and ten thousand to give the answer 10 000.

Week 5: Place value

Resources: cubes in two colours, whiteboards and pens

Vocabulary: place value, place, ones, tens, hundreds, thousands, millions, digit, one-, two- … seven-digit number, tenths, hundredths, thousandths, represents, the same as, equal to, greater, more, larger, less, fewer, smaller, greatest, most, largest, least, fewest, smallest, one … ten … hundred … thousand more or less, compare, order, first, second, third … last, numeral, consecutive

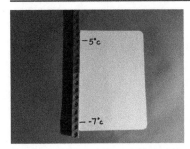

Monday

Give each pair of pupils 20 cubes, 10 each of two different colours. Explain that each cube will represent an interval on a thermometer.

Write –7°C on the board, telling pupils that this was the temperature in Canada one morning. Then explain that the temperature rose by 12°. Ask pupils to create this using the cubes and a whiteboard, changing colours when they reach 0°C. Say the new temperature. Repeat with other start temperatures and increases.

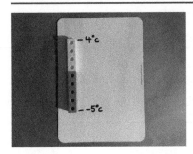

Tuesday

Give each pair of pupils the cubes from Monday's activity.

Write 4°C on the board, telling pupils that this was the temperature in Denmark one morning. Then tell pupils that the temperature dropped by 9°. Ask pupils to create this using the cubes and a whiteboard, changing colours when they reach 0°C. Say the new temperature. Repeat with other start temperatures and temperature decreases.

Wednesday

Give each pair of pupils the cubes from Monday's activity.

Pupils take turns to tell a mathematical story in the context of temperature. Partner 1 in each pair tells the story and partner 2 creates a representation using cubes.

Thursday

Repeat Wednesday's activity, with pupils swapping roles.

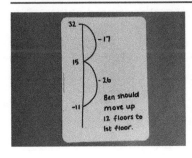

Friday

Give each pupil a whiteboard and pen.

Explain that, when visiting the Empire State Building, Ben gets in the lift at floor 32. The lift goes down 17 floors, and then down a further 26 floors. Ben wants to get out of the lift on the first floor. How many floors does he need to move through and in which direction? (Up 12) Remind pupils that they can draw a number line to check their answer.

Repeat, with other counting backwards / forwards mathematical stories.

Week 6: Place value

Solve number and practical problems that involve place value

Resources: whiteboards and pens

Vocabulary: place value, place, ones, tens, hundreds, thousands, millions, digit, one-, two- … seven-digit number, tenths, hundredths, thousandths, represents, the same as, equal to, greater, more, larger, less, fewer, smaller, greatest, most, largest, least, fewest, smallest, one … ten … hundred … thousand more or less, compare, order, first, second, third … last, numeral, consecutive

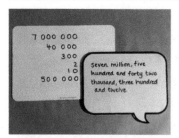

Monday

On the board, write a seven-digit number in expanded form with either the thousands, hundreds, tens or ones value missing as shown. Say the number and ask pupils to write it in expanded form and to show you their whiteboards. Repeat with other numbers.

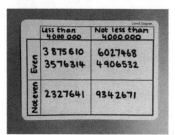

Tuesday

Working individually or in pairs, ask pupils to draw a Carroll diagram on their whiteboards labelled 'Less than 4 000 000', 'Not less than 4 000 000', 'Even' and 'Not even' as shown.

Ask pupils to write the seven-digit numbers shown here in the correct places on their Carroll diagram. Encourage them to jot the numbers down first.

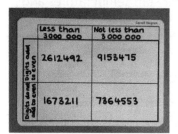

Wednesday

Ask pupils to draw a Carroll diagram on their whiteboards labelled 'Less than 3 000 000', 'Not less than 3 000 000', 'Digits add to even' and 'Digits do not add to even' as shown.

Ask pupils to write the seven-digit numbers shown here in the correct places on their Carroll diagram. Encourage pupils to jot the numbers down first.

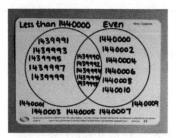

Thursday

Working individually or in pairs, ask pupils to draw a Venn diagram on their whiteboards labelled 'Less than 1 440 000' and 'Even' as shown. Ask pupils to write all the numbers between 1 439 990 and 1 440 010 on the Venn diagram in the correct places.

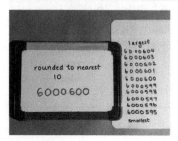

Friday

A number rounded to the nearest ten is 6 000 600. What are the possible numbers? What is the smallest possible number? What is the largest possible number?

Repeat for other problems (e.g. rounded to the nearest hundred or thousand).

Week 1: Algebra

Use simple formulae

Resources: Cuisenaire rods, whiteboards and pens

Vocabulary: place value, place, ones, tens, hundreds, thousands, ten / hundred thousands, millions, digit, one-, two-, three- … seven-digit number, represents, the same as, equal to, greater, more, larger, less, fewer, smaller, greatest, most, largest, least, compare, order, first, second, third … last, numeral, consecutive, estimate, roughly, approximate, exactly, too many / few, round up / down, term, sequence, algebra, express, expression, equation, variable, combination, unknown, possible, formula / formulae, linear

Monday

Explain that this week pupils will be working with formulae. A formula is a type of equation that shows the relationship between different variables.

Give each pair of pupils a set of Cuisenaire rods and allow them to experiment with the rods. Encourage them to compare colours and sizes, creating a staircase with the rods. Ask pupils to talk about the colours and to make simple statements (e.g. light green + purple = black). Pupils record these statements on their whiteboards and feed back to the class.

Tuesday

Give each pair of pupils a set of Cuisenaire rods and a whiteboard and pen.

Partner 1 in each pair writes a simple expression for partner 2 to solve using the rods. Remind pupils that, when they write multiplication sentences, they do not need to use the × sign (e.g. 3 × reds would simply be written as 3 red or $3r$).

Wednesday

Repeat Tuesday's activity, with pupils swapping roles.

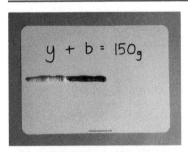

Thursday

Ask pupils to create and write a calculation on their whiteboards to represent the combined mass of two objects. What could the value of y be? And the value of b? Ask pupils to write five different possibilities.

Once pupils are confident, introduce some numbers (e.g. 'The total is 150g. If y is 78g, what is b? If b is 36g, what is y?').

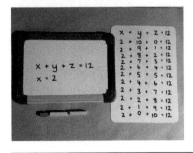

Friday

Give each pair of pupils a set of Cuisenaire rods and a whiteboard and pen.

Write '$x + y + z = 12$' on the board.

Ask pupils to explore all the different combinations for $x + y + z = 12$ when $x = 2$. Encourage and model a systematic method of working.

Repeat with other values for x.

Week 2: Algebra

Generate linear number sequences

Resources: whiteboards and pens, matchsticks / straws / pasta

Vocabulary: place value, place, ones, tens, hundreds, thousands, ten / hundred thousands, millions, digit, one-, two-, three- ..., seven-digit number, represents, the same as, equal to, greater, more, larger, less, fewer, smaller, greatest, most, largest, least, compare, order, first, second, third ... last, numeral, consecutive, estimate, roughly, approximate, exactly, too many / few, round up / down, term, sequence, algebra, express, expression, equation, variable, combination, unknown, possible, formula / formulae, linear

Monday

Explain that this week pupils will be exploring linear sequences: number patterns that increase or decrease by the same amount each time.

Write the sequence on the board as shown. Ask pupils to discuss what happens each time there is a step (a 'term') in the sequence (+7). Repeat with other sequences (e.g. 3, 9, 15, 21, 27 [+6] and 4, 14, 24, 34, 44, 54, 64 [+10]). Encourage use of the vocabulary 'term' and 'sequence'.

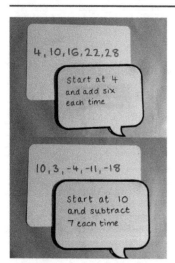

Tuesday

Working in pairs, pupils generate linear number sequences. Partner 1 in each pair generates the sequence. Partner 2 records the sequence, identifying the increase or decrease.

Wednesday

Repeat Tuesday's activity. Today, partner 2 in each pair generates the sequence, including negative numbers. Partner 1 records the sequence, identifying the increase or decrease.

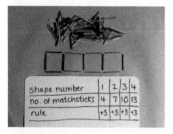

Thursday

Give each pair of pupils a large quantity of matchsticks (or other resource).

Create the pattern shown or draw it on the board. Ask partner 1 in each pair to create the pattern using matchsticks. Partner 2 completes the table on their whiteboard. Demonstrate that the sequence adds 3 matchsticks each time, and that the sequence begins with 4 matchsticks.

Friday

Repeat Thursday's activity, with pupils swapping roles to explore the pattern shown.

Describe linear number sequences

Resources: dice, whiteboards and pens, tape measure / metre stick

Vocabulary: place value, place, ones, tens, hundreds, thousands, ten / hundred thousands, millions, digit, one-, two-, three- ..., seven-digit number, represents, the same as, equal to, greater, more, larger, less, fewer, smaller, greatest, most, largest, least, compare, order, first, second, third ... last, numeral, consecutive, estimate, roughly, approximate, exactly, too many / few, round up / down, term, sequence, algebra, express, expression, equation, variable, combination, unknown, possible, formula / formulae, linear

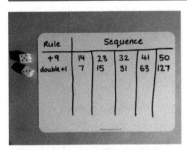

Monday

Explain that this week pupils will be describing linear sequences.

Give each pair of pupils a dice and a whiteboard and pen.

Partner 1 in each pair rolls the dice to generate the start term and partner 2 draws the table shown. Together, the partners complete the table by following these rules for each sequence: rule: +9; rule: double + 1; rule: double – 2.

Tuesday

Give each pupil or pair a metre stick or tape measure and a whiteboard and pen.

Write on the board 77cm, 72cm, 67cm, 62cm, 57cm.

Pupils use the metre stick to find the gaps between each number and use this to describe the sequence (–5).

Wednesday

Give each pair of pupils a whiteboard and pen.

Partner 1 in each pair writes the first four terms of a sequence (with all numbers less than 99). Partner 2 finds the difference between each number, and then writes the next four terms. Together they check their sequence.

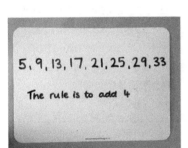

Thursday

Repeat Wednesday's activity, with pupils swapping roles.

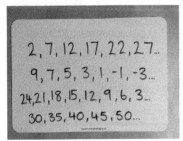

Friday

Call out criteria for problem solving involving linear sequences. For example:

- Generate a linear sequence where the second number is 7. Can you create another one that uses negative numbers?
- Create a linear sequence where the fourth number is 15.
- Write a linear sequence where the rule is 'add 5'.

Week 4: Algebra

Express missing number problems algebraically

Resources: whiteboards and pens, cubes, bowls, playing cards (no picture cards or tens), dice

Vocabulary: place value, place, ones, tens, hundreds, thousands, ten / hundred thousands, millions, digit, one-, two-, three- ..., seven-digit number, represents, the same as, equal to, greater, more, larger, less, fewer, smaller, greatest, most, largest, least, compare, order, first, second, third ... last, numeral, consecutive, estimate, roughly, approximate, exactly, too many / few, round up / down, term, sequence, algebra, express, expression, equation, variable, combination, unknown, possible, formula / formulae, linear

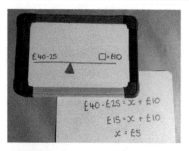

Monday

Explain that this week pupils will be using algebra to describe missing number sentences in the context of equality and balance.

Write the missing number problem shown on the board.

Working in pairs, ask pupils to discuss what the missing number is. Then ask pupils to write the number sentence to match the problem on their whiteboards.

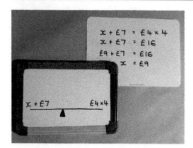

Tuesday

Write the missing number problem shown on the board.

Working individually, pupils write the number sentence to match the problem on their whiteboards.

Model again the stages involved in arriving at the answer for *x*.

Wednesday

Secretly, so partner 2 cannot see, partner 1 takes 2 cards and 2 coloured cubes. They lay the coloured cubes next to the playing cards as shown, attributing the value shown on the card to the coloured cubes. Partner 1 writes the equation that is represented (as shown), providing partner 2 with one of the values. Partner 2 calculates the missing value. Pupils repeat, swapping roles.

Thursday

Give each pair of pupils a bowl / bag of different-coloured cubes and 2 dice.

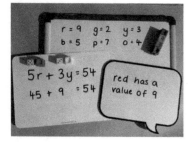

Secretly, so partner 2 cannot see, partner 1 in each pair writes values (between 1 and 9) for each colour (as shown). Partner 2 rolls the two dice and takes 2 cubes from the bag. Partner 1 then writes the algebraic sentence represented by the dice / cube combinations, and the total, and gives one of the colour values. Partner 2 then works out the other colour value. For the next turn, partner 2 takes one new coloured cube, so each turn they find another value of the colours. The challenge is for partner 2 to find all of the values of the colours.

Friday

Repeat Thursday's activity, with pupils swapping roles.

Week 5: Algebra

Find pairs of numbers that satisfy an equation with two unknowns

Resources: string, whiteboards and pens

Vocabulary: place value, place, ones, tens, hundreds, thousands, ten / hundred thousands, millions, digit, one-, two-, three- ..., seven-digit number, represents, the same as, equal to, greater, more, larger, less, fewer, smaller, greatest, most, largest, least, compare, order, first, second, third ... last, numeral, consecutive, estimate, roughly, approximate, exactly, too many / few, round up / down, term, sequence, algebra, express, expression, equation, variable, combination, unknown, possible, formula / formulae, linear

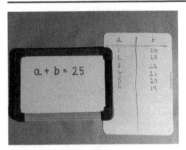

Monday

Explain that this week pupils will be using algebra to explore problems involving whole numbers and measures (length) that have more than one answer.

Write the problem shown on the board. Model the problem with a 25cm length of string and ask pupils to work in pairs to find all possible answers for *a* and *b*.

Encourage a systematic way of exploring all possibilities of *a* and *b*.

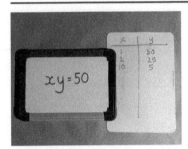

Tuesday

Remind pupils that, when we multiply in algebra, we don't use the multiplication sign. So, *x* × *y* is written as *xy*.

Write the problem shown on the board. Ask pupils to work in pairs to find all possible answers for *x* and *y*.

Encourage a systematic way of exploring all possibilities.

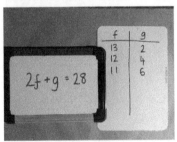

Wednesday

Remind pupils that, when we multiply in algebra, we don't use the multiplication sign. So, 2 × *f* is written as 2*f*.

Write the problem shown on the board. Ask pupils to work in pairs to find all possible answers for *f* and *g*.

Encourage a systematic way of exploring all possibilities.

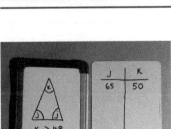

Thursday

Draw an isosceles triangle, as shown on the whiteboard. Explain that *k* is greater than 48, and *j* is a whole number.

What are the possible values for *j* and *k*?

Friday

Remind pupils of the isosceles problem in Thursday's activity.

Ask if *k* will always be an even number when *j* and *k* are whole numbers. Why? Ask pupils to prove it.

Week 6: Algebra

Enumerate possibilities of combinations of two variables

Resources: cubes, whiteboards and pens, objects

Vocabulary: place value, place, ones, tens, hundreds, thousands, ten / hundred thousands, millions, digit, one-, two-, three- ..., seven-digit number, represents, the same as, equal to, greater, more, larger, less, fewer, smaller, greatest, most, largest, least, compare, order, first, second, third ... last, numeral, consecutive, estimate, roughly, approximate, exactly, too many / few, round up / down, term, sequence, algebra, express, expression, equation, variable, combination, unknown, possible, formula / formulae, linear

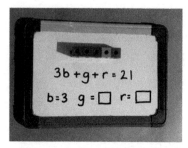

Monday

Use cubes to represent a problem, as shown. Write the algebraic problem alongside.

Reveal the value of one of the colours. Working in pairs and using a whiteboard, pupils calculate the possible values of the other two colours.

Tuesday

Repeat Monday's activity, but with pupils working individually.

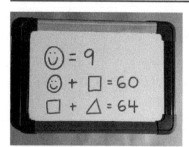

Wednesday

Draw the symbols and values shown on the board.

Ask pupils to use the known value of ☺ to calculate the missing values.

Repeat with other symbols, where one out of the three values is revealed.

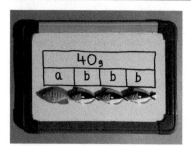

Thursday

On the board, draw the bar model shown to represent the total mass of four objects.

Ask pupils which whole numbers the letters could represent. Encourage a systematic way of exploring all possibilities.

What is the smallest mass *a* can be? What is the largest mass *b* can be?

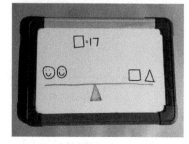

Friday

Draw the problem shown on the board.

Ask pupils to explore possible values for ☺ and the triangle when ☺ is less than 26 and a whole number.

Can the value of the triangle be even? Explain how you know.

Perform mental calculations, including with large numbers

Resources: whiteboards and pens

Vocabulary: digit, add, addition, more, plus, make, sum, total, altogether, one more, two more ... ten more ... one hundred more, how many more to make ...?, missing number, how many more is ...? how much more is ...?, subtract, subtraction, take away, minus, leave, how many are left / left over?, one less, two less ... ten less ... one hundred less, how many fewer?, how much less?, difference between, +, –, =, equals sign, is the same as, boundary, exchange, addend, subtrahend, equivalent, lowest common multiple (LCM)

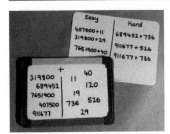

Monday

Ask pupils to split their whiteboards as shown. Write the numbers shown on the board.

Pupils choose a large number and a smaller number from the numbers shown to create three easy and three hard addition sentences.

Pupils perform a mental calculation to solve their questions. Discuss what makes them easy or hard.

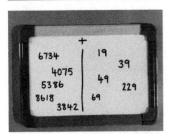

Tuesday

Write the numbers shown on the board. Pupils use the numbers to create ten number sentences that add large numbers plus a near multiple of 10 by compensating. Today's focus is adding near multiples of 10 (e.g. 11, 22, 31, 112, 221).

Demonstrate counting on mentally by adding the multiple of 10, then *adding* the additional 1 or 2 (or adding 1 or 2 and then the multiple of 10). Pupils then answer their own questions.

Wednesday

Repeat Tuesday's activity. Today's focus is adding near multiples of 10 (e.g. 19, 29, 38, 118, 229).

Demonstrate counting on mentally by adding the multiple of 10, then *subtracting* the additional 1 or 2 to compensate. Pupils then answer their own questions.

Thursday

Write the numbers shown on the board. Pupils use the numbers to create ten number sentences that subtract a near multiple of 10 (e.g. 12, 21, 32, 111, 221) from a large number.

Demonstrate subtracting mentally by subtracting the multiple of 10, then *subtracting* the additional 1 or 2 to compensate. Pupils then answer their own questions.

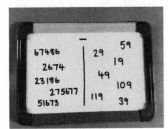

Friday

Repeat Thursday's activity. Today's focus is subtracting near multiples of 10 (e.g. 19, 28, 38, 119, 229).

Demonstrate subtracting mentally by subtracting the multiple of 10, then *adding* the additional 1 or 2 to compensate. Pupils then answer their own questions.

Week 2: Addition and subtraction / Fractions

Add fractions with different denominators using the concept of equivalent fractions

Resources: whiteboards and pens

Vocabulary: digit, add, addition, more, plus, make, sum, total, altogether, one more, two more ... ten more ... one hundred more, how many more to make ...?, missing number, how many more is ...? how much more is ...?, subtract, subtraction, take away, minus, leave, how many are left / left over?, one less, two less ... ten less ... one hundred less, how many fewer?, how much less?, difference between, +, −, =, equals sign, is the same as, boundary, exchange, addend, subtrahend, equivalent, lowest common multiple (LCM)

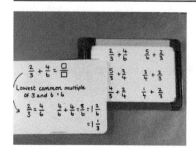

Monday

Explain that before you can add or subtract fractions with different denominators, you must first find equivalent fractions with the same denominator. To do this, demonstrate finding the smallest multiple (known as the *lowest common multiple* or *LCM*) of both numbers.

Pupils practise finding the LCM of fractions shown on the board. They then rewrite the fractions as equivalent fractions with the LCM as the denominator and complete the calculations.

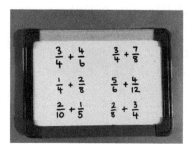

Tuesday

Write addition questions on the board as shown.

Pupils work in pairs to solve the additions. Partner 1 in each pair finds the LCM of both numbers. Partner 2 then rewrites the fractions as equivalent fractions with the LCM as the denominator. Pupils work together to complete the calculations.

Wednesday

Repeat Tuesday's activity, with pupils swapping roles.

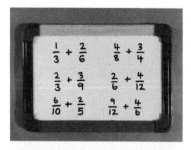

Thursday

Write questions on the board that add fractions with different denominators.

Ask pupils to find the equivalent fractions so that both fractions have the same denominator.

Pupils draw diagrams to represent each calculation, including the answer.

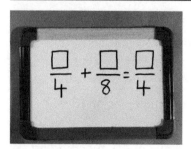

Friday

Write the fraction calculation with missing numbers on the board as shown.

Ask pupils what the missing numbers could be. How many different ways can they find the missing numbers? Pupils could use diagrams to prove their answers.

Add fractions with mixed numbers, using the concept of equivalent fractions

Resources: whiteboards and pens

Vocabulary: digit, add, addition, more, plus, make, sum, total, altogether, one more, two more ... ten more ... one hundred more, how many more to make ...?, missing number, how many more is ...? how much more is ...?, subtract, subtraction, take away, minus, leave, how many are left / left over?, one less, two less ... ten less ... one hundred less, how many fewer?, how much less?, difference between, +, –, =, equals sign, is the same as, boundary, exchange, addend, subtrahend, equivalent, lowest common multiple (LCM)

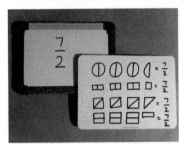

Monday

Write $\frac{7}{2}$ on the board.

Ask pupils to draw as many ways as possible to represent the fraction.

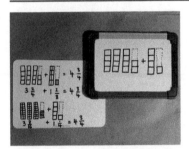

Tuesday

Draw the diagram shown on the board.

Ask pupils to explore as many ways as possible to write the fractions in the calculation, using equivalent fractions and mixed denominators.

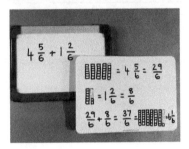

Wednesday

Write the addition number sentence shown on the board. Explain that, to add fractions with mixed numbers, we need to first convert the mixed number to an improper fraction. Model doing so, drawing diagrams to help understanding.

Write other mixed numbers on the board that pupils can practise converting, using diagrams.

Once pupils are confident, you could demonstrate how the whole numbers can be added before the rest is converted.

Thursday

Write a selection of addition problems on the board that use mixed numbers, as shown.

Pupils work in pairs to solve these. Partner 1 in each pair uses drawings to represent their answers. Partner 2 writes their answer on their whiteboard by converting into an improper fraction.

Friday

Repeat Thursday's activity, with pupils swapping roles.

Subtract fractions with different denominators using the concept of equivalent fractions

Resources: whiteboards and pens

Vocabulary: digit, add, addition, more, plus, make, sum, total, altogether, one more, two more … ten more … one hundred more, how many more to make …?, missing number, how many more is …? how much more is …?, subtract, subtraction, take away, minus, leave, how many are left / left over?, one less, two less … ten less … one hundred less, how many fewer?, how much less?, difference between, +, –, =, equals sign, is the same as, boundary, exchange, addend, subtrahend, equivalent, lowest common multiple (LCM)

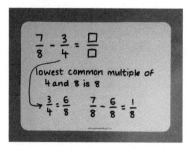

Monday

Explain that, as with addition, before we can subtract fractions with different denominators we must first find equivalent fractions with the same denominator.

Pupils revise and practise finding the smallest multiple (or LCM) of both numbers and completing the calculations.

Tuesday

Write subtraction questions on the board as shown.

Pupils work in pairs to solve the subtractions. Partner 1 in each pair finds the lowest common multiple of both numbers. Partner 2 then rewrites the fractions as equivalent fractions with the LCM as the denominator. Pupils work together to complete the calculations.

Wednesday

Repeat Tuesday's activity, with pupils swapping roles.

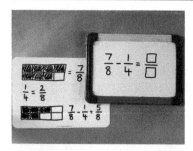

Thursday

Write questions on the board that subtract fractions with different denominators. Ask pupils to find the equivalent fractions so that both fractions have the same denominator.

Pupils draw diagrams to represent each calculation, including the answer.

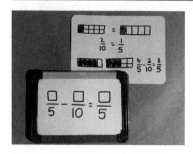

Friday

Write the fraction calculation with missing numbers on the board as shown.

Ask pupils what the missing numbers could be. How many different ways can they find the missing numbers? Pupils could use diagrams to prove their answers.

Week 5: Addition and subtraction / Fractions

Subtract fractions with mixed numbers, using the concept of equivalent fractions

Resources: whiteboards and pens

Vocabulary: digit, add, addition, more, plus, make, sum, total, altogether, one more, two more ... ten more ... one hundred more, how many more to make ...?, missing number, how many more is ...? how much more is ...?, subtract, subtraction, take away, minus, leave, how many are left / left over?, one less, two less ... ten less ... one hundred less, how many fewer?, how much less?, difference between, +, −, =, equals sign, is the same as, boundary, exchange, addend, subtrahend, equivalent, lowest common multiple (LCM)

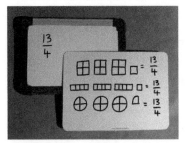

Monday

Write $\frac{13}{4}$ on the board.

Ask pupils to draw as many ways as possible to represent the fraction.

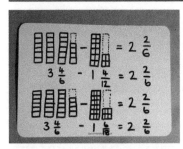

Tuesday

Draw the diagram shown on the board.

Ask pupils to explore as many ways as possible to write the fractions in the calculation, using equivalent fractions and mixed denominators.

Wednesday

Write the subtraction number sentence shown on the board. Explain that, to subtract fractions with mixed numbers, we need to first convert the mixed number to an improper fraction. Model doing so, drawing diagrams to help understanding.

Write other mixed numbers on the board that pupils can practise converting, using diagrams.

Once pupils are confident, you could demonstrate how the whole numbers can be subtracted before the rest is converted ($4\frac{5}{6} - 1\frac{2}{6} = 3\frac{3}{6} = 3\frac{1}{2}$).

Thursday

Write a selection of subtraction problems on the board that use mixed numbers, as shown.

Pupils work in pairs to solve these. Partner 1 in each pair uses drawings to represent their answers. Partner 2 writes their answer on their whiteboard by converting into an improper fraction.

Friday

Repeat Thursday's activity, with pupils swapping roles.

Week 6: Addition and subtraction / Fractions

Perform mental calculations, including with mixed operations

Resources: whiteboards and pens

> **Vocabulary:** digit, add, addition, more, plus, make, sum, total, altogether, one more, two more ... ten more ... one hundred more, how many more to make ...?, missing number, how many more is ...? how much more is ...?, subtract, subtraction, take away, minus, leave, how many are left / left over?, one less, two less ... ten less ... one hundred less, how many fewer?, how much less?, difference between, +, –, =, equals sign, is the same as, equivalent, lowest common multiple (LCM), multiply, divide, find, operation

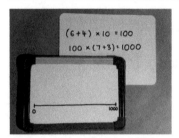

Monday

Write '0 to 1000' on the board. The aim is for pupils to move from 0 to 1000 in the least number of moves using the numbers 2, 3, 4, 6, 7, 10, 40 and 100 and the operations +, –, × and ÷. Explain that each number can only be used once.

How many ways can they find to get from 0 to 1000? Which is the 'quickest'? Can they find a way that uses all the numbers or all the operations?

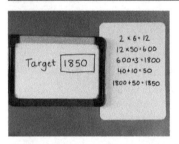

Tuesday

Write 'Target 1850' on the board.

Pupils use the numbers 2, 3, 6, 10, 40, 50 and 100 and the operations +, –, × and ÷ to reach the target. Explain that each number can only be used once. Repeat with other targets.

Wednesday

Write the problem '6348 + 1000 = 8800 – ☐☐☐☐' on the board.

Explain that a digit needs to go in each box to make this calculation correct. Ask pupils to prove it.

Pupils could work individually, but will benefit from working in pairs for this activity to develop their reasoning.

Thursday

Write the problem '5 ☐ 4.31 = 16.55 ☐ 5' on the board.

Explain that +, –, × or ÷ needs to go in each box to make this calculation correct. Ask pupils to prove it.

Pupils will benefit from working in pairs for this activity to develop their reasoning.

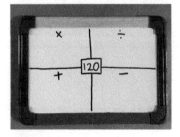

Friday

Draw the grid shown on the board, with 120 in the centre.

Pupils generate a question for each operation where the answer is 120.

Repeat, generating questions where the answer is 123, 499, 4612, etc.

Week 1: Multiplication and division / Fractions

Perform mental calculations, including with mixed operations and large numbers

Resources: whiteboards and pens

> **Vocabulary:** digit, integer, add, addition, more, plus, make, sum, total, altogether, one more, two more ... ten more ... lots, groups of, times, multiply, multiplication, multiplied by, multiple of, product, once, twice, three times ... ten times as (big / long / wide ...), repeated addition, array, row, column, double, halve, share, divide, division, divided by / into, dividend, left, left over, remainder, factors, multiples, prime numbers, lowest / highest common factor / multiple, simplify

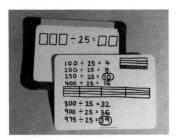

Monday

Write '☐☐☐ ÷ 25 = ☐☐' on the board as shown.

Using whiteboards, ask pupils to explore possible options for this calculation in order to find the largest and smallest dividends without remainders. Encourage the use of diagrams and models to explain how they know they have explored all opportunities.

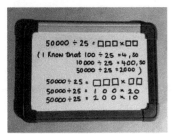

Tuesday

Write '50 000 ÷ 25 = ☐☐☐ × ☐☐' on the board as shown.

Model drawing a quick diagram to represent blocks of 100 as four 25s, then calculating how many hundreds there are in 50 000 to quickly find the answer.

Then ask pupils to write as many different equivalent three-digit by two-digit multiplication number sentences as they can.

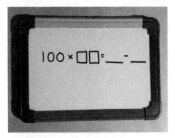

Wednesday

Write '100 × ☐☐ = ____ – ____' on the board as shown.

Ask pupils to suggest any two-digit numbers that could go in the empty boxes. Now ask pupils to write as many subtraction questions as they can where the answer is equal to 100 × ☐☐.

Repeat with other questions: 50 × ☐☐☐ = ___ + ___ and 25 × ☐☐ = ___ – ___.

Thursday

Write 'Target 4851' on the board.

Pupils work in pairs and use the numbers 1, 4, 5, 7, 10, 50, 100 and the operations +, –, × and ÷ to reach the target. Explain that each number can only be used once. Challenge pupils to use all of the numbers (if possible!). Repeat with other targets.

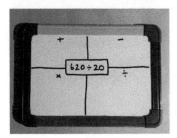

Friday

Draw the grid on the board as shown.

Pupils generate a question for each operation where the answer is equivalent to 620 ÷ 20.

Repeat, generating questions where the answer is equivalent to other large number divisions.

Identify common factors

Resources: whiteboards and pens

Vocabulary: digit, integer, add, addition, more, plus, make, sum, total, altogether, one more, two more … ten more … lots, groups of, times, multiply, multiplication, multiplied by, multiple of, product, once, twice, three times … ten times as (big / long / wide …), repeated addition, array, row, column, double, halve, share, divide, division, divided by / into, dividend, left, left over, remainder, factors, multiples, prime numbers, lowest / highest common factor / multiple, simplify

---

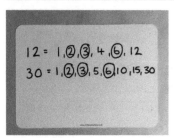

Monday

Explain that *factors* are the numbers you multiply together to get another number (e.g. 2 and 3 are factors of 6). When you find the factors of two or more numbers, some factors are the same and are called *common factors*.

Working in pairs and using whiteboards, ask pupils to find the common factors of 12 and 30, then 48 and 60, and then 15 and 45.

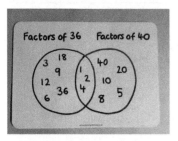

Tuesday

Working individually, ask pupils to draw a Venn diagram on their whiteboards to help them find the common factors of 36 and 40.

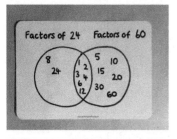

Wednesday

Working in pairs, ask pupils to draw a Venn diagram on their whiteboards to help them find the common factors of 24 and 60.

Demonstrate finding the highest common factor (HCF) of 24 and 60. (12)

Repeat with 36 and 63.

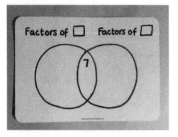

Thursday

Draw the diagram shown on the board.

Tell pupils that you have drawn a Venn diagram to find the common factors of two two-digit numbers. Explain that one of the numbers in the intersection is 7.

Ask pupils what the two two-digit numbers could be. Working individually, ask pupils to explore all possibilities. (The two numbers could be any combination of 14, 21, 28, 35 … up to 98.)

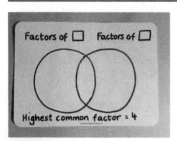

Friday

Draw the diagram shown on the board.

Tell pupils that you have drawn a Venn diagram to find the common factors of two numbers. Explain that both numbers are less than 20 and that the highest common factor is 4. Working individually, ask pupils to find out what the two numbers could be. (8 and 12, and 4 and 16)

Identify common multiples

Resources: whiteboards and pens

> **Vocabulary:** digit, integer, add, addition, more, plus, make, sum, total, altogether, one more, two more … ten more … lots, groups of, times, multiply, multiplication, multiplied by, multiple of, product, once, twice, three times … ten times as (big / long / wide …), repeated addition, array, row, column, double, halve, share, divide, division, divided by / into, dividend, left, left over, remainder, factors, multiples, prime numbers, lowest / highest common factor / multiple, simplify

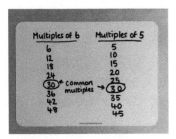

Monday

Explain that a *multiple* is a result of multiplying a number by an *integer* (whole number).

Model finding common multiples of 5 and 6, up to 50, making a list of multiples for each number and then identifying the common multiples.

Ask pupils to work in pairs to find the common multiples of 6 and 9, up to 100.

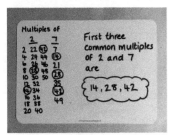

Tuesday

Ask pupils to find the first three common multiples of 2 and 7.

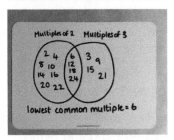

Wednesday

Working in pairs, ask pupils to draw a Venn diagram on their whiteboards to help them find the common multiples of 2 and 3 up to 25.

Demonstrate finding the lowest common multiple of 2 and 3.

Repeat with 36 and 63.

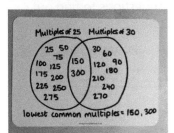

Thursday

Working individually, ask pupils to draw a Venn diagram on their whiteboards to help them find the common multiples of 25 and 30, up to 300.

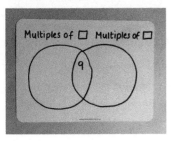

Friday

Draw the diagram shown on the board.

Tell pupils that you have drawn a Venn diagram to find the common multiples of two one-digit numbers. Explain that one of the numbers in the intersection is 9. Ask pupils what the two one-digit numbers could be. Ask pupils to explore all multiples up to 50 for the two numbers.

Identify prime numbers

Resources: 100 squares, sticky notes, whiteboards and pens

Vocabulary: digit, integer, add, addition, more, plus, make, sum, total, altogether, one more, two more ... ten more ... lots, groups of, times, multiply, multiplication, multiplied by, multiple of, product, once, twice, three times ... ten times as (big / long / wide ...), repeated addition, array, row, column, double, halve, share, divide, division, divided by / into, dividend, left, left over, remainder, factors, multiples, prime numbers, lowest / highest common factor / multiple, simplify

Monday

Explain to pupils that a *prime number* is a whole number greater than 1, whose only factors are 1 and itself.

Give each pupil a 100 square. Ask pupils to work from 1–50, highlighting any numbers that do not have factors other than 1 and itself, using a 'Sieve of Eratosthenes'.

(Before they begin, you could ask pupils to predict how many numbers between 1 and 50 they think will be prime numbers.)

Tuesday

Give each pupil a 100 square. Ask pupils to work from 51–100, highlighting any numbers that do not have factors other than 1 and itself.

(Before they begin, you could ask pupils to predict how many numbers between 51 and 100 they think will be prime numbers.)

Wednesday

Give each pair of pupils a 100 square and some sticky notes. Explain that today pupils will be playing 'guess my prime'.

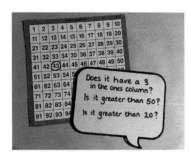

Partner 1 in each pair chooses a prime number between 1 and 100 and writes it on a sticky note. Partner 2 has a 100 square and asks questions to find out which prime number was chosen.

Pupils keep a tally of how many questions they ask. The pupil asking the *fewest* questions wins.

Thursday

Repeat Wednesday's activity, with pupils swapping roles.

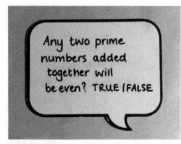

Friday

Set pupils a challenge: 'Any two prime numbers added together will be even. True or false? Prove it!'

(False. Most prime numbers are odd: odd + odd = even. However, 2 is a prime number, so the two prime numbers that you add could be an odd and an even number: odd + even = odd.)

Multiply simple pairs of proper fractions, writing the answer in its simplest form

Resources: whiteboards and pens, squared paper, cubes

Vocabulary: digit, integer, add, addition, more, plus, make, sum, total, altogether, one more, two more … ten more … lots, groups of, times, multiply, multiplication, multiplied by, multiple of, product, once, twice, three times … ten times as (big / long / wide …), repeated addition, array, row, column, double, halve, share, divide, division, divided by / into, dividend, left, left over, remainder, factors, multiples, prime numbers, lowest / highest common factor / multiple, simplify

---

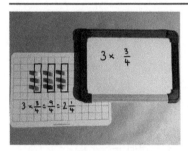

Monday

Give each pupil squared paper / a squared whiteboard and some cubes.

Write $3 \times \frac{3}{4}$ on the board. Pupils draw outlines on squared paper and then place cubes to represent $\frac{3}{4}$, as shown. Describe the multiplication as '3 lots of $\frac{3}{4}$'. Ask how many quarters there are (9). Then use the outlines to calculate the answer and convert it to a mixed number. Repeat for other multiplications of quarters. Ask pupils to write the number sentence underneath.

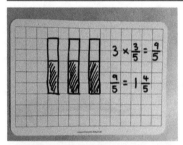

Tuesday

Give each pupil a sheet of squared paper or a squared whiteboard.

Ask them to draw a diagram to represent $3 \times \frac{3}{5}$. Remind pupils that, if the answer is an improper fraction, it will need to be converted to a mixed number. Repeat for other multiplications of fifths, writing the number sentence underneath.

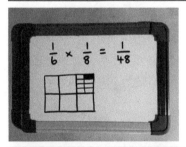

Wednesday

Write '$\frac{1}{6} \times \frac{1}{8} =$' on the board and highlight that these fractions have different denominators. Draw what this multiplication would look like. Make it clear that this is why we multiply the denominators together.

Working in pairs, pupils practise other fraction multiplication questions.

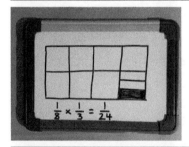

Thursday

Write '$\frac{1}{8} \times \frac{1}{3} =$' on the board and highlight that these fractions have different denominators.

Repeat Wednesday's activity.

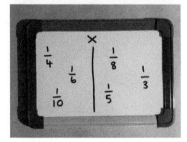

Friday

Draw the grid shown on the board.

Ask pupils to create and solve their own problems using the fractions on the board. Encourage them to use objects or diagrams to support.

Divide proper fractions by whole numbers

Resources: long strips of paper, whiteboards and pens, objects

Vocabulary: digit, integer, add, addition, more, plus, make, sum, total, altogether, one more, two more … ten more … lots, groups of, times, multiply, multiplication, multiplied by, multiple of, product, once, twice, three times … ten times as (big / long / wide …), repeated addition, array, row, column, double, halve, share, divide, division, divided by / into, dividend, left, left over, remainder, factors, multiples, prime numbers, lowest / highest common factor / multiple, simplify

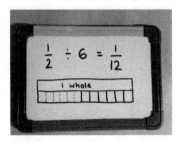

Monday

Give each pupil five long chunky strips of paper.

Ask them to fold and tear one whole strip into quarters. Ask them to divide one of the quarters into three equal parts. Write on the board what pupils have just done: '$\frac{1}{4} \div 3 = \frac{1}{12}$'.

Repeat for other fraction divisions (e.g. $\frac{1}{2} \div 4$, $\frac{2}{3} \div 3$, $\frac{1}{5} \div 4$ and $\frac{4}{6} \div 2$).

Tuesday

Write '$\frac{1}{2} \div 6 =$' on the board.

Draw what this looks like, highlighting the number sentence alongside.

Ask pupils to draw diagrams that represent $\frac{1}{4} \div 3$, $\frac{2}{5} \div 4$ and $\frac{3}{8} \div 2$.

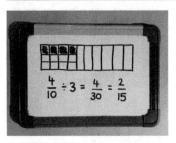

Wednesday

Write '$\frac{4}{10} \div 3 =$' on the board.

Model drawing the calculation, but focus on the written calculation. Ask pupils if they can summarise, without the use of the diagram this time, the process involved in dividing proper fractions by a whole number.

Write more fraction problems for pupils to explore.

Thursday

Draw the grid shown on the board.

Ask pupils to create and solve their own problems using the fractions on the board. Encourage pupils to use objects or diagrams to support.

Friday

Write '$\frac{\square}{\square} \div \square = \frac{4}{12}$' on the board.

Ask pupils to calculate the possible missing numbers.

Week 1: Fractions (including decimals and percentages)

Use common multiples to express fractions in the same denomination

Resources: whiteboards and pens, Cuisenaire rods

> **Vocabulary:** whole, part, equal parts, fraction, one whole, one half, two halves, one quarter, two / three / four quarters, one third, two / three thirds, tenths, hundredths, in every, for every, decimal, decimal fraction / point / place, numerator, denominator, equivalent, same, equal to, factors, multiples, prime numbers, lowest / highest common factor / multiple, simplify

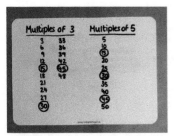

Monday

Remind pupils that a *multiple* is a result of multiplying a number by an *integer*. Model finding the *common multiples* of 3 and 5, up to 50, by making a list of multiples for each number and then highlighting the common multiples.

Ask pupils to work in pairs to find the common multiples of 5 and 8, up to 100.

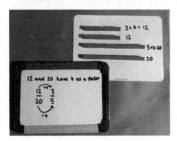

Tuesday

Write $\frac{12}{20}$ on the board. Explain that you are going to simplify this fraction by finding common multiples of 12 and 20 using diagrams and Cuisenaire rods.

Ask pupils what factors there are of 12 and 20 (2 and 4), then use Cuisenaire to explore multiples of 2 and 4 (e.g. 3 × 4 rods are the same length as 12, and 5 × 4 rods are the same length as 20). Demonstrate using 4 as a factor to express the fractions in the same denomination.

Wednesday

Give pairs of pupils Cuisenaire rods and whiteboards and pens.

Write fractions on the board for pupils to simplify, proving the factors using Cuisenaire rods. Pupils will benefit from working in pairs to facilitate verbal reasoning.

Thursday

Give each pupil Cuisenaire rods and a whiteboard and pen.

Write fractions on the board for pupils to simplify, proving the factors using Cuisenaire rods or, if confident, using diagrams. Pupils should work individually now to embed their understanding.

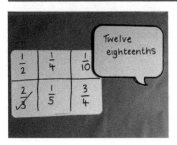

Friday

Ask pupils to draw a 3 × 2 bingo grid on their whiteboards and to write six different fractions of their own choosing in their simplest form.

Call out fractions that have not been simplified. When an equivalent fraction is called out, pupils cross that fraction off their board. The winner is the first pupil to cross out all their simple fractions.

Week 2: Fractions (including decimals and percentages)

Use common factors to simplify fractions

Resources: whiteboards and pens

> **Vocabulary:** whole, part, equal parts, fraction, one whole, one half, two halves, one quarter, two / three / four quarters, one third, two / three thirds, tenths, hundredths, in every, for every, decimal, decimal fraction / point / place, numerator, denominator, equivalent, same, equal to, factors, multiples, prime numbers, lowest / highest common factor / multiple, simplify

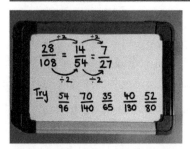

Monday

Write $\frac{28}{108}$ on the board. Explain that large numbers in fraction form make the number difficult to understand and visualise so, wherever possible, we should simplify fractions.

Today pupils will explore simplifying by dividing. To do this we divide both the numerator and denominator of the fraction by the same number (2, 3, 4, 5, etc.) until we can't go any further. Ask pupils to try this for the fractions shown.

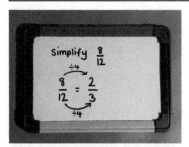

Tuesday

Write $\frac{8}{12}$ on the board.

Explain that another method of simplifying fractions is to find the highest common factor of both the numerator and denominator. Then we divide both the top and the bottom by that factor – in this case, 4. Ask pupils to find the common factors and simplify $\frac{16}{40}, \frac{22}{50}, \frac{18}{48}, \frac{75}{200}$.

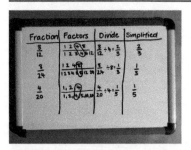

Wednesday

Draw the grid shown on the board. Explain that this could be used as a systematic model of working to find the common factors and the highest common factor. Ask pupils to simplify the fractions shown, using the system demonstrated.

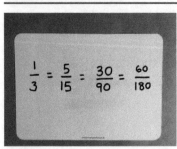

Thursday

A fraction has been simplified three times and is now in its simplest form, $\frac{1}{3}$.

Ask pupils to find three different fractions that it could have been before it was simplified.

Repeat for other fractions (e.g. $\frac{1}{5}, \frac{1}{4}, \frac{3}{4}, \frac{5}{6}$).

Friday

Write '$\frac{1}{2}, \frac{2}{4}, \frac{6}{12}$' on the board.

Ask pupils which of these three fractions has been simplified to its simplest form. Prove it.

Repeat for the other sets of fractions shown.

Week 3: Fractions (including decimals and percentages)

Identify the value of each digit in numbers given to three decimal places

Resources: whiteboards and pens, timer

> **Vocabulary:** whole, part, equal parts, fraction, one whole, one half, two halves, one quarter, two / three / four quarters, one third, two / three thirds, tenths, hundredths, in every, for every, decimal, decimal fraction / point / place, numerator, denominator, equivalent, same, equal to, factors, multiples, prime numbers, lowest / highest common factor / multiple, simplify

Monday

Give each pupil a whiteboard and pen. Ask them to draw a place-value grid as shown.

Start with a variety of five-digit numbers; divide by 10 and describe the changes and the effect of the division. Repeat, multiplying by 10.

Practise multiplying and dividing other five-digit numbers in this way.

Tuesday

Ask pupils to dance or move while you time them for 2 minutes 36 seconds. Record this time on a place-value grid. Then time them dancing or moving for 2 seconds. Explain that 0.236 of a second would be difficult to move for and 0.0236 impossible.

Discuss thousandths in the context of time, explaining how the accuracy is needed in race timings. Now ask pupils to order some race times to show their understanding of the value of the digits.

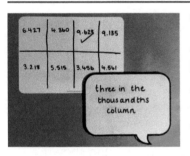

Wednesday

Ask pupils to create eight different one-digit numbers with three decimal places, and to write them in a grid on their whiteboards as shown.

Call out criteria (e.g. 4 in the thousandths column, 2 in the ones, 7 in the tenths), making a note of the criteria. If pupils have a number on their board that matches the criteria, they tick that number. The first pupil to get four in a row wins.

Thursday

Write the digits shown, including a large decimal point, on the board.

Ask pupils how many numbers they can make that are *smaller* than 0.5 using these digits. Ask pupils to convince a partner that one of the numbers they created is smaller than 0.5. They can use a place-value grid, diagrams or other classroom resources.

Friday

Repeat Thursday's activity, with different digits and asking pupils to create numbers that are *greater* than 0.5 but smaller than 1.

Week 4: Fractions (including decimals and percentages)

Multiply numbers by 10, 100 and 1000 giving answers up to three decimal places

Resources: whiteboards and pens, dice

Vocabulary: whole, part, equal parts, fraction, one whole, one half, two halves, one quarter, two / three / four quarters, one third, two / three thirds, tenths, hundredths, in every, for every, decimal, decimal fraction / point / place, numerator, denominator, equivalent, same, equal to, factors, multiples, prime numbers, lowest / highest common factor / multiple, simplify

Monday

Give each pupil a whiteboard and pen and ask them to draw the place-value grid shown.

Pupils practise multiplying the numbers shown by 10 and 100.

Ask pupils to work in pairs to describe what happens to the digits when multiplying by 10 and 100. What is similar and what is different?

Tuesday

Give each pupil a whiteboard and pen and ask them to draw the place-value grid shown.

Pupils practise multiplying the numbers shown by 1000.

Ask pupils to work in pairs to describe what happens to the digits when multiplying by 1000. What is similar and what is different?

Wednesday

Write a six-digit number on the board (e.g. 630 514).

Ask five place-value questions based on this number. Pupils say if they are true or false (e.g. 6305.14 × 10 = 63051.04).

Pupils can then play in pairs using their whiteboards.

Thursday

Give each pupil a whiteboard and pen and ask them to draw the grid shown.

Write a six-digit number on the board and explain that this is the answer. Ask pupils to write the × 10, × 100 and × 1000 fact that would give the answer shown.

Ask what is the same and what is different about the facts they have written.

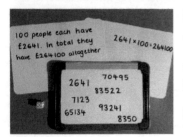

Friday

Write a selection of four- and five-digit numbers on the board.

Give each pair a dice, and a whiteboard and pen. Explain that, when rolled, the dice links to one of these facts: 1 and 4 = × 10, 2 and 5 = × 100, 3 and 6 = × 1000.

Pupil 1 rolls the dice, chooses a number from the board and creates a question. They write the question and answer on their whiteboard. Pupil 2 creates a mathematical story for this question. Repeat with pupils swapping roles.

Divide numbers by 10, 100 and 1000 giving answers up to three decimal places

Resources: whiteboards and pens, dice

Vocabulary: whole, part, equal parts, fraction, one whole, one half, two halves, one quarter, two / three / four quarters, one third, two / three thirds, tenths, hundredths, in every, for every, decimal, decimal fraction / point / place, numerator, denominator, equivalent, same, equal to, factors, multiples, prime numbers, lowest / highest common factor / multiple, simplify

---

Monday

Give each pupil a whiteboard and pen and ask them to draw the place-value grid shown.

Pupils practise dividing the numbers shown by 10 and 100.

Ask pupils to work in pairs to describe what happens to the digits when dividing by 10 and 100. What is similar and what is different?

Tuesday

Give each pupil a whiteboard and pen and ask them to draw the place-value grid shown.

Pupils practise dividing the numbers shown by 1000.

Ask pupils to work in pairs to describe what happens to the digits when dividing by 1000. What is similar and what is different?

Wednesday

Write a six-digit number on the board (e.g. 469 172).

Ask five place-value questions based on this number. Pupils say if they are true or false (e.g. 469 172 ÷ 1000 = 4691.72).

Pupils can then play in pairs using their whiteboards.

Thursday

Give each pupil a whiteboard and pen and ask them to draw the grid shown.

Write on the board a three-digit number with three decimal places and explain that this is the answer. Ask pupils to write the ÷ 10, ÷ 100 and ÷ 1000 fact that would give the answer shown.

Ask what is the same and what is different about the facts they have written.

Friday

Write a selection of three- and four-digit numbers on the board.

Give each pair a dice, and a whiteboard and pen. Explain that, when rolled, the dice links to one of these facts: 1 and 4 = ÷ 10; 2 and 5 = ÷ 100; 3 and 6 = ÷ 1000.

Pupil 1 rolls the dice, chooses a number from the board and creates a question. They write the question and answer on their whiteboard. Pupil 2 creates a mathematical story for this question. Repeat with pupils swapping roles.

Week 6: Fractions (including decimals and percentages)

Multiply one-digit numbers with up to two decimal places by whole numbers

Resources: whiteboards and pens, decimal place-value counters, 1–9 digit cards

Vocabulary: whole, part, equal parts, fraction, one whole, one half, two halves, one quarter, two / three / four quarters, one third, two / three thirds, tenths, hundredths, in every, for every, decimal, decimal fraction / point / place, numerator, denominator, equivalent, same, equal to, factors, multiples, prime numbers, lowest / highest common factor / multiple, simplify

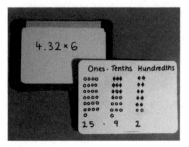

Monday

Give each pair some decimal place-value counters, or ask pupils to represent the counters on a decimal place-value grid. Write a multiplication calculation on the whiteboard containing a one-digit number with up to two decimal places multiplied by a one-digit number.

Pupils calculate the answer on their place-value grid by using their counters / representations. Repeat for other calculations.

Tuesday

Explain that today pupils will practise multiplying a one-digit number with up to two decimal places by a whole number, mentally.

Write 2.3 × 2 on the board.

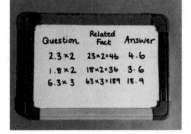

Ask which multiplication fact will help them to answer this (23 × 2). 23 × 2 = 46. Explain that 23 × 2 is 10 times bigger than 2.3 × 2 and ask what we need to do now to use the related fact (divide by 10).

Practise together with other multiplication questions.

Wednesday

Repeat Tuesday's activity, but with pupils working independently to solve problems on their whiteboards, using related facts to support their understanding.

Thursday

Write two equivalent but different multiplication facts on the board as a balance as shown.

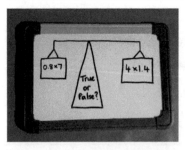

Ask pupils to show you either true or false written on their whiteboards. If true, ask pupils to prove it using a diagram or representation. If false, ask pupils to write an equivalent multiplication calculation on their whiteboards to make it correct. Repeat for other facts.

Friday

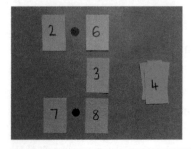

Give each pupil a set of 1–9 digit cards (see back of the book) and two counters.

Ask pupils to use 5 of the cards to make a calculation that fits the form ▢.▢ × ▢ = ▢.▢ (e.g. 2.6 × 3 = 7.8).

How many calculations can pairs find?

0   1   2

3   4   5

6

7

8

9